EPIGENETICS

Brief And Concise Description Of The New Molecular Genetics

Heesum Heesum

Table of contents

Chapter I: Introduction

Definition and Significance of Epigenetics

Epigenetics refers to the study of heritable changes in gene expression or cellular phenotype that do not involve alterations in the underlying DNA sequence. It focuses on the modifications and chemical marks on the DNA and its associated proteins that can influence gene activity and expression without changing the DNA sequence itself.

The significance of epigenetics lies in its ability to provide insights into how the environment and lifestyle factors can impact gene expression and influence disease susceptibility. Epigenetic modifications act as a bridge between genes and the environment, as they can be influenced by factors such as diet, stress, toxins, and even social interactions. By altering gene expression, epigenetic changes can play a crucial role in various biological processes, including development, aging, and disease.

Epigenetics also challenges the long-held notion of genetic determinism, which suggests that our genes alone determine our traits and health outcomes. Instead, epigenetics highlights the intricate interplay between genes and the environment, emphasizing the potential for modifications in gene expression that can be inherited or reversed throughout an individual's lifetime.

Studying epigenetics has broad implications for fields such as medicine, agriculture, and environmental science. In medicine, understanding epigenetic mechanisms can provide insights into the development and progression of diseases like cancer, neurological disorders, and cardiovascular diseases. It can potentially lead to the discovery of novel therapeutic targets and the development of more personalized treatment approaches.

In agriculture, epigenetics research can contribute to crop improvement, disease resistance, and the understanding of plant responses to environmental stresses. In environmental science, epigenetics can shed light on how pollutants and toxins affect organisms and ecosystems, leading to better strategies for environmental protection and preservation.

Overall, the significance of epigenetics lies in its capacity to unravel the complex interactions between genes and the environment, opening up new avenues for understanding health and disease as well as the dynamic nature of biological systems.

A Brief History and Development of Epigenetic Research

The study of epigenetics has a rich history that dates back to the early 20th century. The term "epigenetics" was first coined by embryologist Conrad Waddington in 1942 to describe the interactions between genes and their environment during development.

In the 1970s, scientists discovered DNA methylation, which involves the addition of a methyl group to the DNA molecule, primarily at cytosine residues. This epigenetic modification was found to play a critical role in gene regulation by influencing the binding of transcription factors and other proteins to DNA. DNA methylation became one of the most extensively studied epigenetic marks.

Another important breakthrough came in the 1980s with the identification of histone modifications. Histones are proteins around which DNA is wrapped, forming a structure called chromatin. Modifications, such as acetylation, methylation, and phosphorylation, can occur on the histone proteins and affect gene expression by altering the accessibility of DNA to the transcription machinery.

In the late 1990s and early 2000s, the field of epigenetics witnessed significant advancements with the completion of the Human Genome

Project and the development of high-throughput sequencing technologies. These advancements enabled researchers to explore epigenetic modifications on a genome-wide scale, leading to the discovery of new epigenetic marks and the development of powerful techniques such as bisulfite sequencing and chromatin immunoprecipitation.

The field continued to expand rapidly with the discovery of non-coding RNAs, such as microRNAs and long non-coding RNAs, which were found to regulate gene expression through epigenetic mechanisms. These findings added another layer of complexity to our understanding of gene regulation.

Today, epigenetic research encompasses a wide range of disciplines, including genetics, molecular biology, and biochemistry.

Overview of key concepts and terminology

Before embarking on our exploration of epigenetics, it is essential to familiarize ourselves with key concepts and terminology. We will delve into the mechanisms through which epigenetic modifications occur, including DNA methylation, histone modifications, and non-coding RNAs. We will uncover how these modifications influence gene expression, regulate chromatin structure, and shape cellular identity and function.

Additionally, we will explore the concept of epigenetic programming during early development and the impact of environmental factors on epigenetic modifications. By understanding these fundamental concepts, we can appreciate the pivotal role epigenetics plays in health and disease.

As we embark on this journey into the intricate world of epigenetics, we will delve into various aspects of this rapidly evolving field. We will explore the mechanisms by which epigenetic modifications occur, their influence on development and disease, and their potential for therapeutic interventions. Moreover, we will examine the ethical and social implications of epigenetics and cast our gaze towards the future, exploring emerging technologies and the exciting potential for precision medicine. Join us as we unravel the hidden layers of our genetic regulation and embrace the transformative power of epigenetics in understanding and improving human health.

Chapter II: DNA Methylation

Process and Role in Gene Expression:

DNA methylation is one of the most extensively studied epigenetic mechanisms. It involves the addition of a methyl group (-CH3) to the DNA molecule, primarily at cytosine residues, often occurring in the context of CpG dinucleotides. DNA methylation is catalyzed by enzymes called DNA methyltransferases.

When DNA is methylated, it can influence gene expression in multiple ways. Generally, DNA methylation at gene promoter regions is associated with gene silencing, as it can prevent the binding of transcription factors and other regulatory proteins to DNA. Thus, DNA methylation acts as a repressive mark, inhibiting gene transcription.

Impact on Development and Disease:

DNA methylation plays a crucial role in various biological processes, including development, cellular differentiation, and the maintenance of cell identity. During development, DNA methylation patterns are dynamically regulated, contributing to the establishment of cell-specific gene expression profiles. Aberrant DNA methylation patterns have been associated with numerous diseases, including cancer, neurological disorders, and developmental disorders.

Hypermethylation of tumor suppressor genes can lead to their silencing and contribute to uncontrolled cell growth in cancer. In neurological disorders, alterations in DNA methylation patterns have been linked to changes in gene expression involved in synaptic plasticity and neuronal function. Additionally, environmental factors such as diet, stress, and toxins can influence DNA methylation patterns, further highlighting their role in disease susceptibility.

Histone Modifications

Types of Modifications and Their Effects:

Histone modifications involve the addition or removal of various chemical groups from histone proteins, which are crucial components of chromatin structure. The most studied histone modifications include acetylation, methylation, phosphorylation, and ubiquitination. Each modification has specific effects on gene expression.

For example, histone acetylation, catalyzed by histone acetyltransferases, is generally associated with transcriptional activation. Acetylation neutralizes the positive charge of histones, resulting in a more open chromatin structure and increased accessibility of DNA to the transcription machinery. On the other hand, histone methylation can either activate or repress gene expression, depending on the specific site and degree of methylation.

Epigenetic Regulation of Chromatin Structure:

Histone modifications, collectively known as the histone code, play a crucial role in regulating chromatin structure. By modifying the histone proteins, these epigenetic marks can influence the compaction of DNA into chromatin and the accessibility of genes for transcription.

Histone modifications can recruit specific proteins, such as chromatin remodelers and transcription factors, that either promote or inhibit gene expression. Additionally, different combinations of histone modifications can create a unique epigenetic landscape associated with distinct gene expression patterns, contributing to cell-specific functions and developmental processes.

Non-coding RNAs

microRNAs and Their Regulatory Functions:

MicroRNAs (miRNAs) are short non-coding RNAs that play a vital role in post-transcriptional gene regulation. MiRNAs are transcribed from genomic DNA and processed into mature forms that can bind to messenger RNA (mRNA) molecules. By binding to specific target mRNAs, miRNAs can inhibit translation or induce degradation of the mRNA, thereby reducing the production of the corresponding protein.

MiRNAs can regulate multiple genes simultaneously and are involved in various biological processes, including development, cellular differentiation, and disease pathways.

Long Non-coding RNAs and Their Impact on Gene Expression:

Long non-coding RNAs (lncRNAs) are a diverse group of RNA molecules that do not encode proteins. They are transcribed from specific regions of the genome and can regulate genes.

Chapter III: Epigenetics and Development

A.1 Epigenetic Programming during Early Development:

Imprinting and its Role in Parental Gene Expression

Imprinting is an epigenetic phenomenon that involves the differential regulation of genes based on their parental origin. Imprinted genes are marked with specific epigenetic modifications, such as DNA methylation or histone modifications, during gamete formation. This imprinting pattern is maintained throughout development and can influence gene expression in a parent-of-origin-specific manner. Imprinted genes often play crucial roles in fetal growth, placental development, and brain development. Disruptions in imprinting can lead to developmental disorders and diseases.

Epigenetic Reprogramming and Cellular Differentiation:

During early development, there is a process of epigenetic reprogramming that resets the epigenetic marks inherited from the parents. This reprogramming is necessary for the proper differentiation of cells into various lineages. In mammals, the most extensive reprogramming occurs in primordial germ cells, which give rise to sperm and eggs.

Through a series of demethylation and remethylation events, the previous epigenetic marks are erased and reset to establish a totipotent state in the germ cells. Similarly, during embryonic development, there are waves of global demethylation and remethylation to ensure the proper establishment of lineage-specific epigenetic patterns.

Environmental Factors and Epigenetic Modifications:

B.1 Impact of Nutrition, Stress, and Toxins on Epigenetic Marks:

Environmental factors can influence epigenetic modifications and, subsequently, gene expression during development. Nutrition, for instance, plays a critical role in providing methyl donors such as folate and choline, which are necessary for DNA methylation. Inadequate nutrition during critical periods of development can lead to altered DNA methylation patterns, affecting gene expression and potentially contributing to the risk of diseases later in life.

Stress can also impact epigenetic modifications. Chronic or severe stress has been associated with changes in DNA methylation and histone modifications in various genes, including those involved in the stress response and mental health disorders. Stress-induced epigenetic changes can influence gene expression patterns and affect an individual's susceptibility to stress-related disorders.

Toxic exposures, such as exposure to environmental pollutants or certain drugs, can lead to epigenetic alterations. Chemical compounds can interfere with DNA methylation or histone modification processes, leading to abnormal gene expression patterns. These epigenetic changes induced by toxins can have long-lasting effects on development and increase the risk of various diseases.

B.2: Transgenerational Inheritance of Epigenetic Changes:

Epigenetic modifications can be transmitted across generations, leading to the transgenerational inheritance of epigenetic changes. These changes can occur in both germline and somatic cells. Germline epigenetic changes are inherited through sperm or egg cells and can impact the development and health of future generations. Somatic epigenetic changes, on the other hand, affect non-reproductive cells and can influence an individual's phenotype but are not directly inherited.

Transgenerational epigenetic inheritance has been observed in various organisms, including humans. The transmission of epigenetic marks can occur through several mechanisms, such as altered DNA methylation patterns or histone modifications in germ cells. The environmental exposures of one generation, including nutrition, stress, or toxins, can contribute to these transgenerational epigenetic changes.

Understanding the impact of environmental factors on epigenetic modifications and their transgenerational effects is crucial for elucidating the complex interplay between genes and the environment in development and disease. It highlights the importance of early life experiences and the potential for interventions to mitigate adverse epigenetic changes and promote healthy development.

Chapter IV: Cancer and epigenetic alterations

A.1 DNA Methylation and Histone Modifications in Cancer Development:
Epigenetic alterations play a significant role in cancer development, and dysregulation of DNA methylation and histone modifications is commonly observed in various types of cancer.

Aberrant DNA methylation patterns, characterized by hypermethylation of CpG islands in gene promoter regions, can lead to the silencing of tumor suppressor genes. This can contribute to uncontrolled cell growth and the progression of cancer. Conversely, hypomethylation of specific genomic regions, such as repetitive elements and oncogenes, can promote genomic instability and gene activation, further facilitating tumorigenesis.

Histone modifications also contribute to cancer development. For instance, global loss of histone acetylation, mediated by histone deacetylases (HDACs), is a common feature in cancer. Reduced histone acetylation can result in the compaction of chromatin and transcriptional repression of genes involved in cell cycle regulation and tumor suppression. On the other hand, specific histone modifications, such as methylation or phosphorylation, can lead to altered gene expression patterns, contributing to cancer progression.

A.2 Epigenetic Biomarkers and Targeted Therapies:

Epigenetic alterations in cancer have led to the identification of potential biomarkers for diagnosis, prognosis, and treatment response prediction. DNA methylation patterns and histone modifications can serve as epigenetic biomarkers that reflect the molecular characteristics of tumors.

Epigenetic biomarkers can aid in early cancer detection, the classification of tumor subtypes, and the prediction of patient outcomes. For example, DNA methylation-based tests, such as the detection of methylated DNA in bodily fluids (liquid biopsies), hold promise for non-invasive cancer screening and monitoring. Additionally, specific epigenetic signatures can help identify patients who may benefit from targeted therapies.

Epigenetic-targeted therapies aim to reverse or normalize the epigenetic alterations in cancer cells. Drugs targeting DNA methylation, such as DNA methyltransferase inhibitors (DNMT inhibitors) like azacitidine and decitabine, have been approved for the treatment of certain cancers, such as myelodysplastic syndromes and acute myeloid leukemia.

Similarly, inhibitors of histone deacetylases (HDAC inhibitors), such as vorinostat and romidepsin, have shown efficacy in certain types of lymphoma and multiple myeloma. These targeted therapies aim to restore normal gene expression patterns and reestablish proper cellular control mechanisms.

Neurological Disorders and Epigenetic Dysregulation

B.1 Epigenetic Mechanisms in Alzheimer's, Parkinson's, and Other Conditions:

Epigenetic dysregulation has been implicated in various neurological disorders, including Alzheimer's disease (AD), Parkinson's disease (PD), Huntington's disease (HD), and schizophrenia. Epigenetic modifications, such as DNA methylation, histone modifications, and non-coding RNA-mediated regulation, influence gene expression patterns and contribute to the pathogenesis of these disorders.

In AD, alterations in DNA methylation and histone modifications have been observed in genes associated with neuronal function, amyloid metabolism, and tau pathology. Epigenetic changes can contribute to the dysregulation of genes involved in synaptic plasticity, inflammation, and oxidative stress, which are hallmark features of AD.

In PD, epigenetic modifications influence the expression of genes involved in mitochondrial function, dopamine metabolism, and neuroinflammation. DNA methylation changes in PD-associated genes, as well as alterations in histone acetylation and methylation, have been implicated in the degeneration of dopaminergic neurons.

Chapter V: Epigenetics and Aging

A.1 Epigenetic Changes and Cellular Senescence:

Cellular senescence refers to the state of irreversible cell cycle arrest that cells enter as a response to various stresses and as a part of the aging process. Epigenetic changes play a crucial role in cellular senescence and contribute to aging at the molecular level.

Telomeres, the protective caps at the ends of chromosomes, undergo shortening with each cell division. Telomere shortening is associated with cellular senescence and aging. Epigenetic modifications, such as DNA methylation and histone modifications, can regulate telomere length and telomerase activity. Alterations in these epigenetic marks can accelerate telomere attrition and cellular senescence.

Additionally, senescent cells display specific epigenetic signatures. They exhibit changes in DNA methylation patterns, histone modifications, and chromatin structure. These alterations can lead to the repression of genes associated with cell cycle regulation, DNA repair, and anti-apoptotic pathways, contributing to the senescent phenotype.

A.2 Epigenetic Clocks and Age Prediction:

Epigenetic clocks are mathematical models that use specific epigenetic modifications, primarily DNA methylation patterns, to estimate an individual's biological age. These clocks have been developed based on large-scale epigenomic profiling across various tissues and age groups.

Epigenetic clocks have shown remarkable accuracy in predicting chronological age and have been associated with aging-related processes and diseases. They provide insights into the biological age, which may differ from the chronological age and can be influenced by lifestyle factors, genetic factors, and environmental exposures.

Epigenetic clocks have implications for understanding the aging process, identifying individuals at higher risk for age-related diseases, and assessing the effectiveness of interventions aimed at slowing down aging.

Epigenetic Interventions for Healthy Aging:

B.1 Diet, Lifestyle, and Epigenetic Modifications:

Diet and lifestyle factors have been shown to influence epigenetic modifications and, consequently, the aging process. Nutritional components, such as folate, B vitamins, and polyphenols, act as methyl donors or cofactors for enzymes involved in DNA methylation and histone modifications.

Consumption of a healthy diet rich in fruits, vegetables, and other nutrient-dense foods can support proper epigenetic regulation and contribute to healthy aging.

Physical exercise has also been associated with epigenetic changes linked to beneficial health outcomes. Exercise can influence DNA methylation patterns and histone modifications, leading to altered gene expression related to metabolism, inflammation, and oxidative stress. Regular physical activity has been shown to promote healthy aging and reduce the risk of age-related diseases.

B.2 Potential for Epigenetic Therapies to Slow Aging Processes:

Epigenetic therapies hold promise for modulating the aging process and promoting healthy aging. The manipulation of epigenetic modifications through pharmacological interventions may offer avenues for slowing down age-related changes and mitigating the onset of age-related diseases.

For instance, interventions targeting DNA methylation, such as DNA methyltransferase inhibitors (DNMT inhibitors), have shown potential for rejuvenating aged cells and tissues. These interventions aim to reverse age-associated DNA methylation patterns and restore gene expression profiles reminiscent of younger cells.

Similarly, compounds targeting histone modifications, such as histone deacetylase inhibitors (HDAC inhibitors), have demonstrated effects on cellular senescence and aging. By modulating histone acetylation patterns, these interventions can potentially reverse age-associated changes in gene expression and cellular functions.

However, it is important to note that the field of epigenetic interventions for aging is still in its early stages, and more research is needed to better understand the complexities of epigenetic regulation and its implications for healthy aging.

Chapter VI: Ethical and social implications of epigenetics

A.1 Epigenetics and Personal Responsibility:
The advent of epigenetics raises important questions regarding the balance between genetic determinism and individual agency. Epigenetic modifications can be influenced by various environmental factors, including nutrition, stress, and toxins. This implies that individuals have some degree of control over their epigenetic profiles through their lifestyle choices and exposures.

However, it is crucial to navigate the ethical implications of attributing personal responsibility based on epigenetic influences. Epigenetic changes can occur during critical periods of development or even transgenerationally, which means that individuals may bear the burden of epigenetic changes that are not directly within their control.

Balancing the concepts of personal responsibility, accountability, and the role of environmental factors in shaping epigenetic marks is an ethical challenge.

A.2 Implications for Legal, Ethical, and Social Frameworks:

Epigenetics has implications for legal, ethical, and social frameworks. It raises questions about how to integrate the knowledge of epigenetic influences into existing legal and ethical systems, particularly in areas such as criminal justice, discrimination, and social policies.

In the legal context, epigenetic evidence and its potential to provide insights into an individual's past exposures and experiences could impact criminal responsibility, particularly in cases involving prenatal exposures or early-life adversity. This raises questions about the appropriate use and interpretation of epigenetic data in legal proceedings.

From an ethical standpoint, considerations must be given to issues of privacy, informed consent, and potential stigmatization related to the use of epigenetic information. The disclosure of epigenetic information, particularly in the context of genetic testing or population-level studies, raises concerns about the potential for discrimination, unfair treatment, or the exacerbation of existing health disparities.

Epigenetics and public health

B.1   Epigenetic   Epidemiology   and Population-Level Impact:

Epigenetic epidemiology is a rapidly growing field that examines the relationship between epigenetic modifications and population health. It explores how environmental exposures and social determinants can influence epigenetic marks across populations, potentially leading to differential health outcomes.

Studying epigenetic modifications at the population level can provide insights into the impact of social and environmental factors on health disparities. It can help identify epigenetic signatures associated with specific exposures or diseases and inform public health strategies aimed at prevention, intervention, and reducing health inequities.

B.2 Health Disparities and Epigenetic Factors:

Epigenetics has the potential to shed light on the underlying mechanisms contributing to health disparities. Socioeconomic status, race, ethnicity, and other social determinants of health can influence epigenetic modifications and contribute to health inequalities.

Epigenetic factors may help explain the persistent health disparities observed in populations exposed to socioeconomic disadvantage, discrimination, or adverse environments. By understanding the epigenetic mechanisms involved, interventions can be developed to mitigate the effects of these factors on health outcomes and reduce disparities.

However, the use of epigenetic information to address health disparities also raises ethical concerns. Ensuring equitable access to epigenetic research and interventions requires careful consideration to prevent further marginalization or stigmatization of vulnerable populations.

In conclusion, the ethical and social implications of epigenetics are multifaceted. Striking a balance between genetic determinism and individual agency is essential when considering personal responsibility. Integrating epigenetic knowledge into legal, ethical, and social frameworks requires thoughtful consideration of privacy, informed consent, and potential discrimination. Epigenetic epidemiology provides insights into population health and health disparities, with the potential to inform public health strategies. However, ethical concerns regarding fairness, equity, and the potential for stigmatization must be addressed when utilizing epigenetic information in public health contexts.

Chapter VII: Emerging technologies and methodologies

Advances in epigenome mapping and manipulation:

Epigenome mapping refers to the process of identifying and characterizing epigenetic modifications across the entire genome. Technological advancements have revolutionized our ability to map epigenetic marks, leading to a deeper understanding of their role in gene regulation and disease development.

Some emerging technologies and methodologies in this area include:

Next-generation sequencing (NGS): NGS technologies, such as ChIP-seq (chromatin immunoprecipitation sequencing) and bisulfite sequencing, have significantly enhanced our ability to identify and quantify various epigenetic marks, including DNA methylation, histone modifications, and chromatin accessibility.

High-throughput methods: High-throughput techniques allow the simultaneous profiling of multiple epigenetic marks, providing a more comprehensive view of the epigenome. For example, the development of array-based platforms, such as the Illumina Human Methylation BeadChip, enables large-scale DNA methylation analysis.

Epigenomic editing tools: CRISPR-Cas9-based technologies have opened up new avenues for manipulating epigenetic marks. Epigenome editing tools, such as dCas9-fusion proteins, can be used to target specific epigenetic marks and modulate gene expression. This technology has the potential to uncover the functional consequences of specific epigenetic modifications and provide insights into their therapeutic targeting.

Single-cell epigenetics and spatial epigenomics

Traditionally, epigenetic analyses were performed on bulk cell populations, masking the heterogeneity that exists within tissues. However, recent advances have allowed the investigation of epigenetic patterns at the single-cell level, leading to a better understanding of cellular diversity and dynamics. Additionally, the development of spatial epigenomic techniques has enabled the examination of epigenetic modifications in their tissue context. Key technologies and methodologies in this field include:

Single-cell bisulfite sequencing: This technique enables the assessment of DNA methylation patterns in individual cells, providing insights into cell-to-cell variability and dynamics. It has been instrumental in identifying rare cell populations, characterizing cellular states, and uncovering epigenetic changes during development and disease progression.

Single-cell chromatin accessibility profiling: Techniques like ATAC-seq (an assay for transposase-accessible chromatin using sequencing) and scATAC-seq (single-cell ATAC-seq) allow the investigation of chromatin accessibility at the single-cell level. These methods can identify regulatory regions and infer the activity of transcription factors, shedding light on the regulatory landscape of individual cells.

Spatial transcriptomics and epigenomics: Spatially resolved techniques, such as spatial transcriptomics and spatially resolved chromatin conformation capture (such as Hi-C), enable the examination of gene expression and chromatin interactions in their tissue context. These approaches provide valuable information on how gene expression and chromatin structure vary across different regions within tissues.

Epigenetics and precision medicine

Personalized therapies based on epigenetic profiles

Epigenetic alterations are increasingly recognized as important drivers of disease and present promising targets for personalized therapies. Understanding an individual's epigenetic profile can aid in the identification of specific therapeutic targets and guide treatment decisions. Some key aspects of personalized therapies based on epigenetic profiles include:

Biomarker discovery: identifying specific epigenetic modifications or signatures that are associated with disease progression, prognosis, or response to treatment is crucial for developing personalized therapeutic strategies.

Integrating multi-omics data, including epigenomic, transcriptomic, and clinical information, can facilitate the discovery of robust biomarkers.

Epigenetic drug development: the development of drugs that target specific epigenetic.

Chapter VIII: Conclusions and Future Perspectives in Epigenetic Research

Recap of Key Insights and Findings in Epigenetic Research:

Throughout this exploration of epigenetic research, several key insights and findings have emerged, highlighting the significance of epigenetics in understanding gene regulation, development, disease etiology, and therapeutic interventions. The key points to recapitulate include:

Epigenetic Modifications: Epigenetic modifications, such as DNA methylation, histone modifications, and non-coding RNAs, play critical roles in gene expression and cellular identity. They regulate various biological processes, including development, aging, and response to environmental stimuli.

Transgenerational Epigenetic Inheritance: Epigenetic marks can be inherited across generations, impacting the phenotype and disease susceptibility of offspring. This phenomenon emphasizes the importance of considering not only genetic but also epigenetic factors in understanding heritability and intergenerational health outcomes.

Epigenetics and Disease: Epigenetic alterations are implicated in a wide range of diseases, including cancer, neurological disorders, cardiovascular diseases, and metabolic disorders. They can serve as diagnostic biomarkers, therapeutic targets, and indicators of treatment response.

Environmental Influences: Environmental factors, such as nutrition, stress, toxins, and lifestyle choices, can modulate epigenetic marks, leading to long-lasting effects on health and disease susceptibility. Understanding gene-environment interactions through epigenetics provides insights into disease prevention and personalized interventions.

Reflection on the Potential of Epigenetics in Understanding and Improving Human Health:

The research presented in this work underscores the immense potential of epigenetics in unraveling the complexities of human health and disease. By elucidating the epigenetic mechanisms underlying gene regulation, we gain a deeper understanding of the molecular basis of diseases and the interplay between genetic and environmental factors. Key reflections on the potential of epigenetics include:

Precision Medicine: Epigenetic profiling can aid in the development of personalized therapies by enabling targeted interventions based on individual epigenetic profiles. This approach holds the promise of enhanced treatment efficacy and reduced adverse effects, leading to improved patient outcomes.

Early Disease Detection and Prevention: Epigenetic biomarkers can provide valuable tools for early disease detection, risk assessment, and monitoring. They offer the potential to identify at-risk individuals, enabling timely interventions and preventive measures to mitigate disease progression.

Therapeutic Interventions: Epigenetic modifiers, including DNA methyltransferase inhibitors and histone deacetylase inhibitors, have shown promise as therapeutic agents in various diseases. Further exploration and refinement of epigenetic drugs could lead to more effective treatment options and novel therapeutic strategies.

Environmental Health: Understanding the impact of environmental factors on the epigenome can inform public health policies and interventions to mitigate the adverse effects of environmental exposures. Epigenetic research provides a basis for promoting healthier lifestyles, reducing exposures to harmful agents, and optimizing environmental conditions.

Call to Action for Further Exploration and Application of Epigenetic Knowledge:

While significant progress has been made in epigenetic research, there are still numerous avenues to explore and challenges to overcome. To fully harness the potential of epigenetics in understanding and improving human health, a call to action is necessary:

Continued Research: The exploration of novel epigenetic modifications, the development of advanced methodologies, and the integration of multi-omics data are essential for further unraveling the complexities of epigenetics. Continued research efforts are crucial for expanding our knowledge and translating findings into clinical applications.

Collaborative Efforts: Epigenetic research requires interdisciplinary collaborations among scientists, clinicians, computational biologists, and bioethicists. Collaborative initiatives can facilitate data sharing, standardization of protocols, and the establishment of comprehensive databases to gain a better research on epigenetics.

Printed in Great Britain
by Amazon